Hanging out with
GRAVITY

Gary McCallister
and
Zane McCallister

CONTENTS

DEDICATION

Help! I've fallen in love and I can't get out.
Thanks for believing in me, Gaydra!
Gary

To my wonderful Father without whom I wouldn't exist.
Zane
(Actually, I wrote this part! – Gary)

ACKNOWLEDGMENTS
Gary -

I wasn't so much raised by my parents as I just grew up, sort of like a weed. I will never be able to thank them enough for that benign neglect. Time provided me with opportunities to hunt, fish, hike, explore, build, take apart, read, dream, write, and in general enjoy a wasted youth. These experiences later became critical to my education, both in the humanities and science.

My first, major, academic influence was Mrs. May Robinson who taught me honor's English as a freshman in college. She opened up the world of ideas and academia for me.

Dr. Ferron Andersen at Brigham Young University accepted me, much later, as a graduate student with very little evidence that it was a wise move. As he expressed it at the time, "Well, Gary, I believe there are such things as late bloomers." Without doubt, he was the single greatest influence in my choosing to become a scientist.

I must acknowledge the generosity of Dr. Gerald Schmidt at the University of Northern Colorado for enabling me to complete my doctoral program under his instruction. He was a leading parasitologist of his day, and his influence was enormous.

I also want to thank my son Zane McCallister, who helped develop these exercises in science education. Zane is co-author of this book, and he reminds me that he gets to choose "my" home someday.

Zane –

I wasn't so much raised by my parents as I just grew up, sort of like a weed. I will never be able to thank them enough for that benign neglect. It provided me with opportunities to hunt, fish, hike, explore, build, take apart, read, dream, write, raise critters, catch pigeons, play sports, date a lot of girls, and in general enjoy a wasted youth. This later became critical to my education, both in the humanities and science.

I also want to thank my father, Gary McCallister, who helped develop these exercises in science education. He is co-author of this book, and he occassionally gets reminded that Iwill eventually get to choose his "senior" home.

WHAT A WIERD WAY TO ORGANIZE A BOOK

When a child is born, the brain knows nothing. If you tell a toddler not to touch something because it is hot, that child possibly understand "No" and "touch". What the child doesn't understand is "hot". Your admonition will almost guarantee that the child will touch what it is that's hot sooner rather than later.

The body instructs the mind. There is no way to understand hot and cold until you have experienced them. The same is true of forward and back, right and left, right and wrong, and almost any concept you can think of. Only after you experience the reality, can you talk about the idea.

This is so true it is impossible to speak of abstract ideas without using physical adjectives. Colors are not only colors but they are loud, soft, muted, or sharp. Consider the big ideas of looking forward, a backward person, tall orders, political left or right, under handed, short shrift, widespread, bright ideas, little consequence, low opinions, heavy news, weighty opinions, and dramatic asides.

Adults usually surmise that the mind controls the body. That's why schools spend so much time teaching ideas, and so little time teaching students how to actually do things. They assume the thought precedes the action. In truth, the experience must precede the thought.

That is why in this book, the game, experiment, experience, activity, and behaviors are presented first in Part 1. Part 2 consists of a quasi-scientific, semi-serious, partially historical, occasionally-text-bookish, and a generally unfortunate presentation of some of the ideas and known facts concerning gravity.

You are welcome to do the activities first or last. You may also read in between times or skip any portion of either section. Why would I care? I already understand gravity. You are the one who seems to want to know more. So, if you want to read, turn to Part 2. If you want to have some fun and do something, start with Part 1.

Hanging out with
GRAVITY
Part One

I haven't always paid close attention to gravity. It's kind of a mundane subject. Gravity just always seems to be hanging around, and I take it for granted. . . . . (Hanging around! Get it? If it weren't for gravity there would be no hanging around. Everything would be floating around is my guess.)

Anyway, if gravity has got you down (Okay, sorry! I'll quit.) this little **game** will turn your world upside down. Of course, up is relative in terms of gravity. In fact, up and down are two very significant concepts.

In our world, there are just three dimensions, and they can be defined by the following directions: forward, back, left, up and down. For now we'll investigate up and down. Together they represent a third all dimensions.

Down is significant because that is where we fall. But falling has all kinds of interesting characteristics. How we fall depends on mass, weight, force, acceleration, velocity, friction, inertia, and all kinds of other things that we don't

When NASA first started sending up astronauts, they quickly discovered that ballpoint pens would not work in zero gravity. To combat the problem, NASA scientists spent a decade and $12 billion to develop a pen that writes in zero gravity, upside down, underwater, on almost any surface including glass and at temperatures ranging from below freezing to 300C.

The Russians used a pencil.

usually think about while we are actually falling.

The experience of falling can actually be quite invigorating if one doesn't worry too much about the act of stopping the fall. Stopping a fall is a fascinating concept in itself. The cientific term for this is "ouch".

The words highlighted on the previous page will probably show up again. You might want to remember them.

You probably already know a lot about falling, and stopping a fall, just from walking around, swinging in swings, jumping on trampolines, and falling out of bed. Wouldn't it be nice, though, if we could predict these events a little more carefully? Maybe we could even use gravity to help us accomplish tasks like moving heavy machinery, or shooting accurately. We don't really need a book to tell us how to fall, or how to stop falling. What we need is experience and a way to analyze our experiences. We need MARBLES!

Physics Teacher: "Isaac Newton was sitting under a tree when an apple fell on his head, and he discovered gravity. Isn't that wonderful?"

Student: "Yes, Sir! If he had just been sitting in class looking at books like us, he wouldn't have discovered anything."

Science isn't about facts, books, and memorizing stuff. Science is actually a way of learning about things by observing specific objects or events and trying to understand the patterns or order so we can better predict or control those physical objects or events.

In the game we are going to play, observe the effects of gravity, so you can understand it better. In this game you will become a scientist by:

- making observations,
- forming theories,
- testing theories (through experimentation),
- evaluating your theories and reforming them as necessary,
- developing new concepts about gravity, and
- improving your ability to predict the effects of gravity.

WATCH CAREFULLY

Let's make an observation. That is what scientists do more than anything else. They just sit around and watch what is happening around them. Sometimes the things they look at are wierd. Take Sir Isaac Newton. The apple didn't really fall on his head, you know. However, he was watching an apple fall when he got curious.

But who stands around watching an apple until it falls? Didn't he have a job or something to do? Maybe he just saw it happen out of the corner of his eye. You have to admit, though, that if there were no gravity, there would be no real reason for anything, coming detached from a tree, to necessarily fall "down" to the earth.

From Newton's observation, he eventually thought up his first law. Newton's First Law is that every object that is in a state of rest tends to remain in that state unless an external force is applied to it. (Yeah, they used to talk that way.) But you can test this for yourself.

Put a marble on the floor in front of you. It may roll a little depending on the slope of your floor. (That, by the way, is exactly what we're going to explore). When it comes to rest, watch it! Watch it, closely! Observe carefully! See the marble! See the student. See the student watch the marble. Is the marble watching the student? How long can the student watch the marble?

I think you will eventually decide that a marble at rest will remain at rest until an external force is applied. You are now one third as smart as Sir Isaac Newton.

GAME ON!

The Playing Field

This **game** can be played wherever anything, or anyone, is falling down. A skateboard park would be ideal. However, as you will see, there are certain risks involved in falling which can sometimes interfere with our ability to make careful observations, mental reflection, calculation and pattern discernment. You may want to start on a smaller scale, both for safety and ease of understanding.

> *We learn nothing until something changes. Even if we can't move the thing we want to examine, we move our own position in relation to it.*

Materials Needed:

If you want to start small you will need to gather together the following items.

1. two ramps of equal height
2. several marbles (two of each size and kind)
3. two tape measures
4. a plastic drinking cup, cut in half longitudinally (One cup provides two halves.)
5. masking tape
6. a semi-smooth, long surface

Science has a reputation for being precise, accurate, and all that stuff. Sometimes it is. However, experiments often start out fairly crudely. (That's my kind of science!) It's a lot more fun, and cheaper too. Of course, cheaper doesn't impress the funding agencies. But it might impress a parent or a spouse. Actually, a lot of science was discovered with very cheap stuff.

Many different items will work for this experiment. You may have noticed that they are not precisely defined. For a ramp I have often used a 2x4 block of wood to prop up a length of quarter round wood molding. I cut a notch in the 2x4 so the quarter round will fit in it. The height of the ramp is not significant as long as the two stands are equal.

The length of the quarter round is not significant either. But again, the two should be equal in length. I usually make mine about 30 cm long. A marble is going to "fall down" this ramp, so any piece of wood, or other material, will work as long as it has sides to keep the marble falling in a straight line.

The figure below illustrates how each "race track" is to be set up. Set up stand, ramp, tape measure and cups as shown.

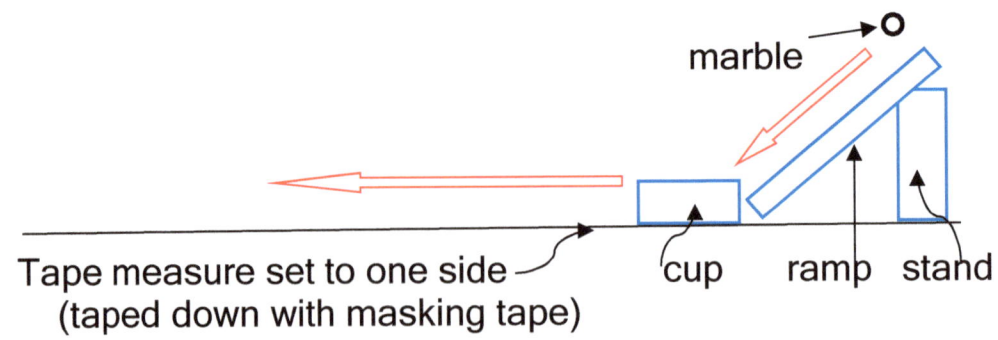

marble

Tape measure set to one side
(taped down with masking tape)

cup   ramp   stand

RING AROUND THE ROSIES
POCKET FULL OF POSIES
ASHES, ASHES
*WE ALL **FALL** DOWN*

FALLING DOWN IS SOMETIMES FUN!

Gravity is not just a good idea.
It's the law!

THE HISTORY OF FALLING

-1	0	1	6 AD?	1400 BC
Satan falls from Heaven	Adam falls for Eve	Eve falls for Satan's lie	Icarus falls to death on wax wings	Jericho falls to noise pollution

The Rules of the **Game**

1. The **game** consists of letting a marble fall down the ramp in such a way that it rolls into the plastic cup and causes the cup to slide across the floor.
2. Players are issued playing equipment and are allowed several minutes to conduct test runs. In this way they familiarize themselves with the events.
3. A moderator announces how far the cup should be moved by the released marble. The players, or teams of players, try to position the ramp and marble in such a way as to move the cup exactly that distance.
4. A trial run is conducted, and the actual movement of the cup is measured. The difference between the actual cup movement and the target distance is recorded.
5. Three trials are run for three different target distances.
6. At the end of the three trial runs, and after measuring the three differences in actual cup movement compared to target distances, the amounts of error are added up for the three trials. The lowest score wins since the player or team with the lowest error came the closest to predicting successfully.
7. During play, the players may set their ramp at any angle they choose, place the marble at any position on the ramp for release, and use any of the marbles in their possesion to achieve the result.
8. All measurements should be calculated using the metric system.
9. Players should do all of their own measurements, perform their own calculations and record their own results. (Of course, an appropriate monitor can aid with difficult determinations, accuracy, and honesty. Trust but verify. Ronald Reagan said that.)

Players may not lose their marbles.

> *Contrary to popular belief, science is not an individual pursuit. It is almost always done in teams and with collaborators. So you might have more fun playing this game with more than two players. The number of players do not even need to be the same. In many ways having several team members makes for lively argumentation and learning.*

WHO IS THE BEST "FALLER"?

	Target	Team 1	Team 2	Difference
Trial 1				
Trial 2				
Trial 3				
Average				
Total				

Work sheets, similar to the one above for running contests, can be found at the end of this book. (These may, of course, be duplicated as needed for continuing experiments and refinements.)

We find, in this mundane world, that knowledge and accomplishment, in and of themselves, are sometimes not sufficient motivation for activity and learning. Prizes may be awarded, at the moderators preference, for best performance However, do not use any of these as winners prizes:

- all expense paid trips to anywhere
- all you can eat of anything
- any kind of special privilege
- drums, harmonicas and banjos are advised against
- 43 oz. sugary drinks and/or any form of caffeine
- Cars, motorcycles, or the latest Nike running shoe
- Puppies, kittens, hamsters, goldfish, reptiles or parakeets

Science is absolutely dependent on the honesty of scientists. If a scientist reports false data, it leads to incorrect conclusions. There are times when incorrect conclusion can be deadly, such as in calculating data needed to keep an airplane airborne. Incorrect conclusions also cause others to formulate false hypotheses and waste a tremendous amount of time, energy and resources in attempting to solve problems. Scientists who falsify data are shunned by the scientific community.

SCIENCE BEGINS

When you don't know why something happens, you formulate an idea in your mind about the cause. Suppose you find yourself forced to sit in your lonely room for something you have done. It may not be entirely clear to you what behavior caused this situation.

- Was it because you took the cookie?
- Was it because you hit a sibling?
- Was it because you lied?
- Or was it because you hollered and screamed?

> *Science is a great game. It is inspiring and refreshing. The playing field is the universe itself.*
> *Isidor Isaac Rabi*
> *Nobel Prize 1944.*

Without knowing the "cause" of what happened, you would be reduced to repeating each of these behaviors, one at a time, to determine which is or is not an "out of control" offense. Several different explanations are plausible. However, if you were to repeat them all at once, with the same results, you would still not know which behavior was the critical one.

However, if you were to select one behavior and repeat it, carefully noting the outcome, you can then, by the process of elimination, discover which behavior is most likely to get you sent to your room. There are some flaws in this analogy. But, in general, this would be considered the "scientific approach".

THE HISTORY OF FALLING (continued)

1617	1618	1877	1901
Faust Vrancik first successful parachute jump	Phillip Fabricius survives being thrown from a window	"Zazel" becomes first human cannon ball	Annie Tyler 1st to survive Niagara Falls in a barrel

You may have used this already without realizing how

sophisticated you were being!

When we select a given explanation for an event, we have formed an *hypothesis*. Whoa! Wait a minute. That's all there is to it? You make a guess about cause and effect and it's called a hypothesis. That's the same as a theory. I'm a scientist!

It is now time for each player, or team, to form a hypothesis about how the marble rolls, how the cup slides, how the cookie crumbles . . . (No wait! That's something else.) Anyway, the hypothesis will help the team better predict the outcome of each event, so they can perform better on the next game.

Sometimes it is useful to write out a hypothesis. It helps to see exactly what you are thinking. A good way to do this is to use an "If . . . Then . . . ." statement. "If I increase the ramp angle, then the

The hypothesis might include changing some of the following:
- the angle of the ramp
- where you place the marble on the ramp to start
- which marble to use

You may want to write out your hypothesis for later reference. In fact, you may want to add notes and calculations during your game playing here. It would be a good idea to obtain some kind of notebook for keeping notes and calculations in. You will almost for sure want them later.

Just like the "sitting-in-your-room" analogy, you might want to test things separately . . . . Wait a minute. Let's not get ahead of ourselves. Testing hypotheses is the next step in our game.

"The trouble with having an open mind, of course, is that people will insist on coming along and trying to put things in it."
Terry Pratchett

THINKING ONE (and creating)

Have you ever noticed that if you are reading or writing and someone needs to speak to you, they always apologize for interrupting? Sometimes they might even wait until later to talk to you as they can see you are busy. However, if you are lost in deep thought they never even apologize.

I find this odd, because often when I am reading or writing, I can return to my task and know where I left off. It is almost impossible to return to where I was thinking and find my place again. Don't you think that is strange? We are thinking something in our brain but if we are interrupted we can't recall where we were in our own brain. This makes it far more important to be able to think uninterrupted.

So I have found this trick. If you make it look like you are doing something while you are thinking, people tend to leave you alone more to think.

So how can you think about gravity, marbles, ramps and cups without looking like you are thinking? Here is one way. Create a work of art that depicts something you know or think about gravity. This might be a drawing, painting, water color, sculpture, tapestry, or any other art form. This makes you look busy, but you're really thinking. Shhhh. Don't worry, I won't rat you out.

See, gravity is invisible. It is a force or an attraction, but it can't be seen. It's like wind like that. We can see the effects of wind, but we can't see the wind. So just thinking about how to make something that is invisible visible will require that you think a lot about whatever it is that's invisible that you want to see. Like gravity.

Oh, I   already know what you're going to say. "I'm not artistic!" "I don't know how to do those things!" IT DOESN'T MATTER!!!! The point is to LOOK like you're doing something when you are really thinking. It keeps

People sometimes are confused about the differences between art and science. Personally, I am just confused. However, if you are too, perhaps it will help you to think about the two fields in this way.

people out of your hair when you're doing important stuff like thinking. No one will care what it looks like. It is just what you thought of to think about gravity.

OK, here's an attempt I made to create art work around gravity. That's me, as a boy, climbing an aspen tree, upside down.

You will want to go ahead and create some art work now, even though you may feel it is silly. It may prove important at the International World Series and Championship of the Marble Game

EXPERIMENTS

When one does an experiment and everything works properly, it is boring. What that means is that we already understood the object or phenomenon we were experimenting with.

When our results do not meet our expectations, then it gets interesting, because we know that we do not really understand what is going on.

Remember hypotheses? You could try an experiment. As soon as you are allowed your freedom, pick one of the behaviors that you hypothesize got you into trouble, march right out and repeat it. Whack your brother a good one, or help yourself to the cookies. It doesn't matter.

Just pick one. This is called an experiment. Careful observation following the experiment to see what the outcome is will help explain which behaviors are acceptable and which are not.

Remember that marble you put on the floor and watched to verify Newton's first law of gravity? Well, put it back. Make sure it is obeying the first law, or call 911 if it refuses and request backup. If it does settle down after a while, make sure it is still and then flick it with your finger.

Isn't that fascinating? The marble moved! You have just verified Newton's Second Law: for every action there is an equal and opposite reaction. Of course, we haven't proved that the reaction was equal, but the marble did move. At least we demonstrated the action/reaction part.

You are now two thirds as smart as Newton. Oh, all, right, I suppose you are just 10/16ths as smart as Newton. (This would be so much easier in metric.)

(Back to EXPERIMENTS and gravity)

Now let's apply the idea of an experiment to your gravity problem. Your ideas about how to move the cup more accurately might have included changing some of the following:

- The angle of the ramp
- Where you place the marble on the
- ramp to start
- Which marble you use

However, if you try to test all of them at once you won't know the effect of any given change. So you might want to see what happens when you just change one thing at a time, like the angle of the ramp.
When you keep everything else the same, such as the same marble for each trial, you are doing an experiment.

Robert Hooke, an irascible, but brilliant, English scientist, wrote a letter to his friend Isaac Newton. Isaac Newton was the one who started all this trouble with gravity. Prior to him everyone just fell down and forgot about.

Anyway, Robert Hooke wrote this letter and kind of one-upped Isaac. He calculated what would happen if one dug a hole clear through the earth, from one side to the other. The he speculated that you dropped a "gravity train" through the hole.

(A "gravity train is a hypothetical construction and has no similarity to a "gravy train". He was speculating on this as a mode of transportation to lower shipping costs, I guess.)

Anyway, he calculated that the 'train would acerbate to 25,000 miles per hour, until it reached the center of the earth, when it would begin to decelerate until it came to a complete rest on the other side. The entire trip he estimated would take forty two minutes and twelve seconds. However, they wouldn't serve meals or snacks.

Now if you are serious about experimenting with all these variables, you will notice right away that there is also a lot of variability in results. The same marble, at the same location on the ramp, with the
ramp at the same angle will often move the cup slightly different distances.

There is a chance that this only happens to me when I do experiments. If that is the case then it is a function of my, err, uhm, variability. However, if you should discover the same thing you should not look upon your data as a reflection of your poor technique. Instead you should shrug and blame your little bother, or something.

It would be good to write down some of your results. This way, during your next bout you can refer to these notes to improve your interesting thing about data, interesting thing besides the fact that it can help you pulverize the competition in this stupid game, is that it is predictions. When you write down results of experiments those results are called data. They are plural.

If you run five tests with the middle sized marble and write down your results you will have data. If you run only one test and write it down it would properly be called datum. So at the dinner table you would say please pass the data if there were more than one on a plate, and please pass the datum if there were only one left.

ROUND TWO

Round two is played exactly the same as round one. The moderator calls out a distance the cup is to be moved and each team chooses the angle of the ramp, length and marble they think will best achieve that result. The actual results are recorded and the difference between the goal and the achievement is calculated. The lowest score after three trials is the winner.

The difference is that now you are two thirds as smart as Newton, and you have data to help you make predictions.

Did you do better than last time?

Winning isn't everything. Unless the prize is big enough I guess. What I mean is, even if you didn't win, if you did better than last time it means you understand gravity a little better than

> Take Stephen Peer for example. In 1887 he attempted to walk across Niagara Falls, at midnight, on a steel cable only five eighths of an inch thick. Understandably he had a few drinks before starting out. This man obviously understood very little about gravity.
>
> To somewhat offset his ignorance, there is a theory that he didn't misstep but was shot by a funambulatory* rival. Unfortunately his body was never found so this is not confirmed.

you did before. That is no small matter. It could be the difference between a trip and a fall.

We are so accustomed to gravity that we have a hard time understanding that people didn't always know about gravity.

*Funambulatory is not a physics word, although in my opinion it should be. It means: 1. Pertaining to rope-walking. 2. Narrow, like the walk of a ropedancer. 3. Performing like a rope dancer

As in: "tread softly and circumspectly in this funambulatory track and narrow path of goodness: pursue virtue virtuously: leaven not good actions nor render virtues

Anyway, can you think of new ways to improve your performance? Are there new experiments you could? Are there ways to might improve without experimentation?

Here are some questions you might ask yourself.

- Have you weighed your marbles?
- Is there a relationship between weight and distance?
- Have you weighed your cup?
- Is there a relationship between cup and marble weight?
- What is the difference between weight and mass?
- Have you measured the angle of your ramp?
- Is there a relationship between angle and movement?
- Have you tried placing the marble at different distances up the ramp?
- Why would distance up the ramp make any difference?
- Could you use fewer variables?
- Could you calculate from mass, distance, and angle instead of running numerous trials?

Dang! Why does it always come back to thinking? I hate

that!

THINKING TWO

Okay, so what can I do so people will leave me alone while I think? I know. You could write a story or poem about gravity, or where gravity is a main player.

<u>*I really think you will want to do this!*</u>

Listen, I'm not going to say any more about it,
but I am very sure you will want to stop
here now and do this.

THINKING THREE

> The word "mass" comes from a Greek word that refers to a "barley cake", or lump of dough. I think this is where we get the phrase "to amass a fortune" from a confused understanding of what exactly dough is.

There is a subtle difference between the mass and weight. In physics, mass can be defined as an object's resistance to acceleration. For example, if I am standing very still on the edge of a terrifying height, I have a mass of, for *purely hypothetical reasons,* let's say 300 pounds. It will require an effort equal to 300 pounds to make be move from my position. This resistance to movement is called inertia. My wife just calls it stubborn.

This is especially evident on a flat floor. Pushing a 300 pound box across a flat floor would require a great deal of energy. (I am ignoring friction for now, and you should too. Because I said so, that's why!)

So what is weight? Weight is a different, yet related, property of matter. It is the gravitational force acting on a given body. Gravity is the attraction between two physical bodies and is related to the sizes of the bodies and the distance between them. So if the two bodies differ in size, or if they are farther apart, the effect can vary.

Mass is an intrinsic measure of an object that doesn't vary. Weight can vary with size and distance. The obvious example is that you weigh a different amount on the moon than you do on earth. Your mass remains the same.

> "Humor is the only test of gravity, and gravity of humor; for a subject which will not bear raillery is suspicious, and a jest which will not bear serious examination is false wit."
>
> Aristotle

That is why things are more tenuous when you are standing on a high cliff. Your mass is the same, but as soon as you lean a little, gravity, the thing that attracts you to the earth, begins to exert a pull which lessens the amount of actual force needed to overcome inertia. So a slight disturbance in your position could have amplified effects.

What effect do you suppose your marbles mass has on its roll down the ramp? What is interesting is that the marbles weight (response to gravity) is the force that must overcome the cups inertia (resistance to acceleration).

THE HISTORY OF FALLING (continued)

1972	1974	1987	2008
Vesna Vulovic survives a 33,316 fall from a DC 9	Queen Elizabeth visits Pentecost Island to witness bungee jumping (falling) to ensure	Study published on 132 cats falling from high rise buildings	Father Adelir de Carli went aloft

In 1000 helium ballons - balloons recovered

In 1987 the Journal of the American Veterinary Medical Association published a study based on 132 cats that fell from high rise buildings. I don't think they actually dropped the cats. I mean, they're Veterinarians. I think they just analyzed data from published reports.

Interestingly, it seems that the higher the fall, the better the survival rate. Their conclusion is that in falls from more than seven stories the cat was able to contort its body in order to maximize surface area and create a small parachute effect that lessened their speed. I think there is a moral to that data, but I'm not sure what it is.

CALCULATING

Sir Isaac Newton said this:

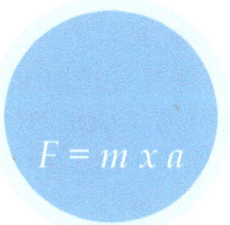

$$F = m \, x \, a$$

where m = mass and a = acceleration.

According to Newton this cat fall data is all wrong. He claimed that everything falls at exactly the same rate in relation to the earth's gravity. He calculated that the speed at which one falls under gravity will increase by approximately 9.8 meters per second, every second we are allowed to fall. In the first second an object would fall 9.8 meters, but in the next second we would fall 19.6 m/s, and so on. This only becomes a significant issue if you fall for more than one second, so it's alright.

> Calculating is another form of thinking. Most of the time people think with words. But when we think with data we sometimes have to think in different ways.

So anyway, that's how fast your marble should be falling. Can you calculate the marbles speed and determine if it is falling correctly? Cats apparently don't fall correctly.

How fast should [the cat be] traveling after one [second?] How fast after two [seconds?] How fast after three [seconds?] If a story of a building [is ten feet] high, how fast will the [cat be] moving after 70 feet?

> The word stems from calculus, which, today, is seen as a form of inhuman punishment. However, originally the word comes from the Latin calx, which referred to a small stone in the gall bladder.

> Calx also meant a small pebble, which was the first way that people used to count. Later they strung the pebbles into an abacus which was a kind of "calculator" used by ancient cultures.

Anyway, calculating is cool because it is another way to do something while you are thinking so that people don't interrupt your thinking. So long as you have a bunch of numbers on the page or screen, people know you are thinking hard and won't bother you. You do have to appear as if you are thinking, however. Numbers on a page do not disguise a nap.

Anyway, you are now fully as smart as Newton, having just learned his third law of gravity: the relationship between an object's mass (M), its acceleration (a) and the applied force (F) is: $F = MA$. However, can you now use this information to better help you play the game?

> This particular version of the marble game is not played for scientific purposes. It's played by loading the top with marbles, which, when released cascade down the ramps in a terrible clatter until they all fall out the bottom. A dexterous child can often load the marbles back into the top as fast as they fall out making a cacophony of noise that can last nearly indefinitely. In fact, science has yet to find out how long it

PS – what is the formula for acceleration (a)? Could you calculate the mass of your marble using information about the cup distance? How?

ROUND THREE

Amazing! The game never changes, just the players and the skill level.

Did you do better than last time?

Here is an interesting ==iteration== of the game. Have the moderator announce a single distances in advance and then allow each player or team to discuss, think and calculate for an extended period of time prior to a sudden death bout. The only change in this game is that the player or team cannot have access to the equipment for experimentation prior to the event.

A lot of life is iteration. First you walk holding onto something, then you walk unassisted, then you run, and then skip, etc. It may not feel as good as winning, but iteration is a necessary step to winning.

Pentagon iteration. Connecting alternate corners of a regular pentagon produces a pentagram which encloses a smaller inverted pentagon. Iterating the process produces a sequence of nested pentagons and pentagrams.

TAKE A BREAK

Can you write a song or music inspired by gravity? I don't

often think of gravity in a musical sense. But all that thinking and calculating in one way has been kind of tiring. Try and think about gravity in another way.

- What if there was no gravity?
- How does gravity create beauty?
- Can gravity be made visible?
- Are the effects of gravity meaningful?
- What is gravity an analogy for?
- What do the words gravity and grave have in common?
- What is the opposite of gravity?
- What does gravity sound like?
- How is gravity like love, or beauty, or children, or sunsets?

Why don't you see what you can do with writing music or a song about gravity? It will be a nice break **AND YOU WILL REALLY BE GLAD YOU DID IT.**

Did I mention that you will really be proud of this down the road!

Iteration means the act of repeating a process usually with the aim of approaching a desired goal or target or result. Each repetition of the process is also called an "iteration," and the results of one iteration are used as the starting point for the next iteration.

THINKING (again?!) THREE

The Dependence of Traffic Ticket Cost on Automobile Speed

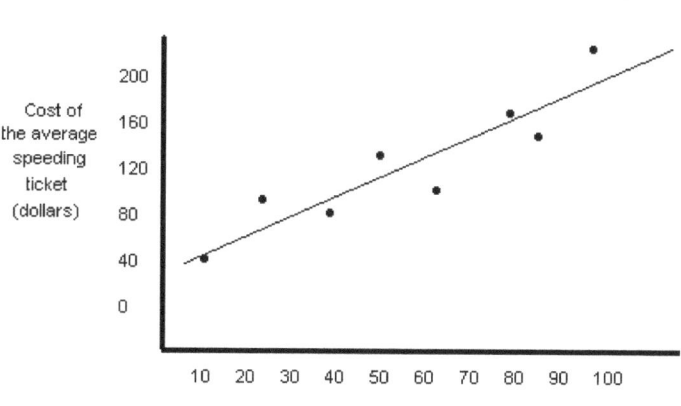

Multiple Choice Question:
If I drive 100 miles per hour, how much do you think I will be fined if I get pulled over?
a. $40.00
b. $160.00
c. $120.00
d. $200.00

The clarity of this graph and the way it helps me make a relatively accurate prediction is almost stunning. I wonder if I could use something like this to help me predict how far the cup will move. Hmmmmm!

The trouble, of course, is that every trial is a little different. My cup doesn't move the same distance each time, even when I keep all the variables steady.

Hey, here's a thought. What if I ran multiple trials with the same marble, in the same position, and with the same ramp angle and then used the average to create my graph. You know how to do that. Just add up all the distances and then divide by the number of distances. The result is the average (statisticians sometimes call this the "mean"). Such a graph might be more accurate under real world conditions involving dusty floors and uneven surfaces, than calculations of an ideal theoretical result.

More than one line can be drawn on a graph. You can either label the line, or draw the line with different appearances: dotted lines, dashed lines, colored lines.

ROUND FOUR

Don't peak too soon.

Here is a terrifying thought. Half the people in the world are below average.

Preparing for the
INTERNATIONAL MARBLE GAME OLYMPIAD

Okay, loosen up! Shake it out! The time you have been waiting for is approaching. What? You didn't know you were waiting for a time? Well, let me tell you, you were! The great, the only (I'm pretty sure it's the only), the magnificent, International Marble Game Olympiad will soon be here.

This is no time to tense up. Just like an athlete, you have got to relax and back off the training the last little while if you want to break through that barrier and peak at the right time. There are any number of activities you can do to prepare.

In the next little while, before the big day, prepare your final portfolio and relax a little. By now, you have learned a number of different ways of thinking, analyzing, calculating and predicting. Now would be a good time to organize your thoughts, theories and calculations in a way that would be easy to access.

Part of the final competition is your technical notebook in which you show your calculations, trial data, thoughts and results. This data book, lab book, field book, is a part of the scientific game. Yours will be public record, just like a real scientist.

You have also created art work, poetry or stories, and music about gravity. Get them all together. Make a display. Be able to explain your endeavor, techniques, design, how it portrays and represents gravity. These will be part of the competition.

You are allowed five displays in the competition:
- Your technical data log
- Your art work
- Your writing
- Your music

Of course, you have only done three so far. Your technical data will be your fourth display.

You can spend your preparation time doing one final project of your choosing. Here are some suggestions, but you are free to tackle almost any topic you are interested in, as long as it relates to gravity.
- Creating a list of internet sites about gravity
- Making a gravity game
- Reading books on gravity
- Collecting art work, lyrics, music or other references to gravity
- Trajectory and armaments: archery, guns, artillery, catapults
- Build a trebuchet or catapult
- Research gravity based businesses or activities
  - Zip lines
  - Flight
  - Mountain climbing
  - Swimming

Finally you will want to make an appointment with your monitor and plan the big events. You will want to consider many of the items located in the sidebar on this page.

This is a great "falling down" competition! Be sure it is well planned and conducted with the full "gravity" which it deserves (which may not be much). This may be the only contest in the world where those who fall best, win. Good luck and God bless!

Activities and displays you might want to consider for an International Marble Game Olympiad.

- o *Readings*
- o *Music*
- o *Contests*
- o *Bump and jump*
- o *Trampoline*
- o *Egg drops*
- o *Music chairs (fall down)*
- o *Lectures on various topics (experts or participants)*
- o *Invite the press*
- o *Display areas for contestant projects*
- o *judges*

WORKSHEETS FOR THE FALLING GAME
WHO IS THE BEST "FALLER"?

	Target	Team 1	Team 2	Difference
Trial 1				
Trial 2				
Trial 3				
Average				
Total				

WHO IS THE BEST "FALLER"?

	Target	Team 1	Team 2	Difference
Trial 1				
Trial 2				
Trial 3				
Average				
Total				

WHO IS THE BEST "FALLER"?

	Target	Team 1	Team 2	Difference
Trial 1				
Trial 2				
Trial 3				
Average				
Total				

	Target	Team 1	Team 2	Difference
Trial 1				
Trial 2				
Trial 3				
Average				
Total				

WHO IS THE BEST "FALLER"?

	Target	Team 1	Team 2	Difference
Trial 1				
Trial 2				
Trial 3				
Average				
Total				

WHO IS THE BEST "FALLER"?

	Target	Team 1	Team 2	Difference
Trial 1				
Trial 2				
Trial 3				
Average				
Total				

WHO IS THE BEST "FALLER"?

	Target	Team 1	Team 2	Difference
Trial 1				
Trial 2				
Trial 3				
Average				
Total				

WHO IS THE BEST "FALLER"?

	Target	Team 1	Team 2	Difference
Trial 1				
Trial 2				
Trial 3				
Average				
Total				

Hanging out with
GRAVITY
Part Two

They try to tell me that gravity is a natural phenomenon by which all physical bodies attract each other. I don't see anything natural about it at all except that it exists. I think it's downright spooky. It is most commonly recognized as the force that causes physical objects to fall toward the ground when dropped from a height. I have never experienced anything other than things falling down, but if you think about it, why don't things just float off into space. It's weird.

It's also supposed to be the thing that gives us weight. Like I need that! Here all this time I thought it was because I ate too much. If this is true I can at least now deny responsibility for my weight and blame it all on the size of the earth.

Gravity is supposed to one of the four fundamental forces of nature. The others are electromagnetism, and something called the nuclear strong force and the nuclear weak force. Scientists are so creative. Interestingly, none of these can be seen directly but are assumed to exist from indirect evidence. This makes them very much like faith or prayer, except we know the rules better.

Gravitation is a force of attraction that acts between, and on, all physical objects with matter (mass) or energy. Gravity is the only force acting on all particles with mass; it has an infinite range; it is always attractive and never repulsive; and it cannot be absorbed, transformed, or shielded against. Those are pretty interesting properties not shared by other fundamental forces.

GALILEO

Galileo was a very religious man and attended church faithfully. I know people like to tell the tale about how he was persecuted by the church for his gravitational theories. Actually that is new age propaganda. The church actually tried him for what I suppose we would call breach of contract and dishonesty. But you can read that story elsewhere if you really want the truth. Just don't read Bertolt Brecht's version. It's not true.

Anyway, one Sabbath morning during mass he was watching the chandelier after it had been pulled to one side to light the candles. When it was released it swung to and fro like a pendulum. He noticed that while the length of the arc of the swing grew smaller slowly until it stopped, the time it took to complete the arc remained the same.

See, I could have done that. Almost anyone could. Once I wrote a story that got published and a friend said, "I could have written that." She was probably right. However, she didn't. It's all about what we actually do, not what we could have done. That means of course that people need to do things.

So Galileo did something. That seems to be the difference between Galileo and me.) He went home and tied a rock to a string and tried to mimic the same actions with a controlled pendulum. He could have written a grant request to the National Science Foundation and got about a billion dollars for a new, up-to-date, cutting- edge pendulum lab but the NSF didn't exist in the 16th and 17th centuries.

In fact, Galileo didn't even have a cell phone or a digital clock and had to rely on his pulse to measure the time of each swing. But he was right. The distance of the arc (period) of the swing grew shorter and shorter but the time remained the same. He tried playing around with different sized rocks and different lengths of string. He discovered that the period grew longer as the string grew longer. But the weight had nothing to do with the period. Things seemed to fall independently of the weight, or put another way, all things fell at the same rate regardless of how much they weighed.

Now you might not think that what Galileo did was so hard. I suppose it wasn't in a way. It's just that until then humans hadn't done anything. (Well, OK they did a lot of things, but they didn't do silly things like playing with pendulums.) Prior to Galileo mankind pretty much just accepted the proclamations of Aristotle that things of different weight fell at different rates. Nobody really went out and checked things out.

Many people refused to believe this new hypothesis and Galileo went on to a number of other experiments and calculations to prove his point. That's is where the story about dropping things off the leaning Tower of Pizza comes from, although, like the tory of his church persecution, it is probably not completely true.

What we do know is that Galileo did go on to employee inclined boards and rolling marbles (oh, all right, he used rolling bars) extremely similar to the exercises in this book, to flesh out his gravitational theories. I am sure he would have used cut in half plastic Dixie Cups had they been available in his day. Our version of Galileo's experiments have been modified to include new concepts such as mass, acceleration, and momentum. That makes it more complicated and therefore more interesting and fun to run as a contest.

By using an elongated incline plane Galileo slowed the descent of the marble so it could be observed. He still didn't have a digital clock, stop watch, or cell phone so he used a water clock. This is a device that allows water to drip through a spigot that can be turned on and off. Time is then measured by weighing the amount of water. Lack of equipment has never hampered a real scientist.

You can exactly duplicate Galileo's experiment by using a flat board that is 6 feet long and propped up two inches high on one end. The slope of the board will be its height (2 inches) divided by the number of inches long it is (6X12). That equals 1/36. That means that anything rolling down the ramp is rolling at $1/36^{th}$ the speed by which it would fall straight down. Another way of saying this is that gravity will be reduced by $1/36^{th}$.

If you don't have a water clock handy, you can use almost anything with a regular interval. A metronome works well. This is small machine that ticks the time loudly. It is used, although not often enough, by music students to get the correct timing of their music. Mark the position of the marble after each second for about four or five seconds.

You will discover that the distance traveled at the end of two seconds will be about four times the distance traveled in one second. After three seconds the marble will have traveled nine times as far as the first second. After the fourth second the marble will have traveled 16 times as far, and after the fifth second the marble will have traversed 25 times as far as the first second.

Now, I might have noticed the swaying chandelier. I might even have noticed the period and the time. But to tell you the truth, the numbers 4, 9, 16, and 25 look almost random to me. Not to Galileo. What he noticed was that these numbers were the squares of the time traveled. ($2^2=3$, $3^2=9$, $4^2=16$, and $5^2=25$)

Next he did the experiment with heavier and lighter weights and discovered that the time intervals remained the same. In other words, he decided that the distance covered increases as the square of the time for all objects. Everything falls at the same rate.

The moral of this story is that you should pay attention in Church because God might reveal something important to you.

NEWTON

Apples have been associated with falling since the beginning of time. After all it was the apple that led to the fall of Adam and Eve. The apple of Paris started the Trojan War that led to the fall of Troy, William Tells apple fell off the head when it was shot with an arrow and helped establish one of the earth's most stable countries. Then there is the story of an apple falling on Newton's head.

The law of universal gravitation was postulated by Sir Isaac Newton. Newton actually did get his ideas from an apple although, apparently, the apple didn't hit him on the head. He told many people that he was out wandering through a garden when he saw an apple fall. It apparently just struck him one day, the idea, not the apple, that something had to be pulling the apple towards the earth.

It has always made me wonder a little about Newton. I mean, didn't he have a job to do or something? Who has time to wander in gardens? Was he actually watching apples fall, or did he just sort of see it happen out of the corner of his eye? Anyway, there are two great lessons to be learned from his experience besides the fact that Newton was a strange fellow.

First, this story demonstrates the value of wandering around outdoors in gardens instead of going to school. If he'd been in class he would never have discovered anything. It also shows how science is first and foremost a matter of asking strange questions.

It's not as if people hadn't already had rich experiences with gravity from just walking around, tripping, swinging from vines in the trees, and falling out of bed. There were plenty of observations, I suppose. What was needed was for someone to ask, "Why do things all fall in the same direction?" His answer was to declare that there must be some sort of invisible force that acted over distance to attract things to it. At the time, some people accused him of dabbling in the occult. I mean, really? An invisible force? You have to admit it sounds suspicious.

Newton asked himself what would happen if he stood on a mountain and fired a bullet parallel to the ground. The bullet would have two motions: horizontal with the velocity imparted by the gun, and falling with the force of gravity. If the earth was flat the bullet would eventually strike the earth.

However, since the earth is not flat but curved, the bullet will be pulled toward the earth by gravity along a curved path. If the velocity is sufficient (and there was no friction from the atmosphere) the bullet might never strike the earth but continue in orbit around the earth. Voila! A satellite.

From this he reasoned that if the moon was attracted to the earth by gravity it was perpetually falling to earth. However, if the velocity of the fall was sufficient it would forever fall, and forever miss the mark.

So his real contribution though was to ask, "How high does gravity go?" Newton decided it had to go quite a ways into the air because the apple tree was pretty tall. This led him to speculate that perhaps this invisible force extended all the way to the moon! Even he was surprised to calculate that it apparently went all the way to "infinity and beyond", to quote Buzz Lightyear. In fact, his theory and calculations, more or less, settled the question of how and why the moon circles the earth and the planets circle the sun.

Anyway, Newton did not receive any compensation for his most famous slogan: "A body at rest tends to remain at rest unless an external force is applied to it." That is only one of his three wise sayings. Knowing this makes you now one third as smart as Isaac Newton.

Unfortunately he never explained how to avoid those external forces. I am working on that problem by spending as much time as possible remaining at rest in the garden. You might say that gravity has kind of got me down.

Newton published his book on gravity, entitled, Principia, in 1687. In this book he hypothesized the inverse-square law of universal gravitation. The kind of spoke differently back in those days. Would you like to hear how he said it?

"I deduced that the forces which keep the planets in their orbs must [be] reciprocally as the squares of their distances from the centers about which they revolve: and thereby compared the force requisite to keep the Moon in her Orb with the force of gravity at the surface of the Earth; and found them answer pretty nearly."

I am not exactly sure what all that means, but it works pretty well for some things. Newton's theory was used to predict the existence of the planet Neptune, based on the motion of Uranus, long before it was actually observed. Newton himself didn't predict this, but others using Newton's theory calculated its position and subsequent discovery.

It's ironic that it was a discrepancy in a planets orbit that led to the knowledge that Newton's laws had some flaws.

A discrepancy in Mercury's orbit pointed out flaws in Newton's theory. By the end of the 19th century, it was known that its orbit showed slight perturbations that could not be accounted for entirely under Newton's theory, but all searches for another perturbing body (such as a planet orbiting the Sun even closer than Mercury) had been fruitless. The issue was resolved in 1915 by Albert Einstein's new theory of general relativity, which accounted for the small discrepancy in Mercury's orbit.

Although Newton's theory has been superseded, most modern non-relativistic gravitational calculations are still made using Newton's theory because it is a much simpler theory to work with than general relativity, and gives sufficiently accurate results for most applications involving sufficiently small masses, speeds and energies.

ENERGY

These two assumptions provides an accurate approximation for most physical situations. I must admit however, I have never calculated any of that for myself. I have had a lot of experience with gravity before I ever even heard of Newton, and I just mostly go on what experiences has taught me. For example, gravity can be used to exert force. It is a form of energy itself. We usually can't see energy any more than we can see gravity, but you have probably felt its affects. Of both gravity and energy I mean.

If you put a heavy box on the top shelf of the closet, you have created potential energy. It's called potential energy because the box isn't really doing anything. But the box has the potential to fall from the shelf due to gravity, severely injuring your foot. There is no need for you to do this experiment. Just take my word for it.

When the box falls, it is no longer considered potential anything because it is now actually doing something, namely falling. The movement of the box is called kinetic energy. Kinetic is defined as, "of, or relating to, the movement of physical bodies." I will now use the word in a sentence. My grandchildren are very kinetic.

Physicists tell me that the amount of energy released when a box falls is exactly equal to the amount of energy that was required to lift the stupid box onto the shelf in the first place. This is one reason why I don't trust physicists very much. The chaos, damage, and pain created by the falling box obviously exceeded the energy required to lift it. But if you can suspend disbelief for a few minutes, I think I can show you why this idea is extremely important.

Let's suppose that you lift one of your kitchen table chairs onto the table. It now could potentially fall down. Next lift a second chair and place it on top of the first chair. Which chair has more potential to fall? The second one of course, because you had to lift it higher than the first chair, which took more energy. Besides, the higher the stack the more unstable it is. Now lift a third chair on top of the other two. Now what has more energy?

This is a bit of a trick question since you will guess it is the third chair for the reasons just explained. In fact, your wife will probably be the most energized as she realizes the threat to her orderly household! Disregarding your wife, it would be the third chair. Disregarding your wife, however, would not be wise. Perhaps you should do all this when she is not home.

The point is that the more things that are stacked on each other, the more energy the top part of the resultant stack represents. What if we didn't use kitchen chairs but instead stacked shoe boxes? The amount of energy in the stacked shoe boxes would be less, sure. But the third box would still possess the most energy of all the boxes.

What if we stacked nickels? The top nickel would represent the most energy, and the more nickels in the stack the more energy. Even if you stack playing cards the act of stacking would still represent the storing of energy. This may not seem very significant to you because even if a playing card fell on your foot it wouldn't hurt much.

Furthermore, the concept doesn't change when we pick up a tiny particle like a hydrogen atom and stack it onto a carbon atom. It is really no different from stacking a kitchen chair on the table, except that your wife probably won't get all mad about it. You will have still created potential energy. If you stack more than one atom onto the carbon atom, you will have created more energy than if you stack only one. A stack of hydrogen atoms on a stack of carbon atoms is otherwise known as sugar.

Now, if you knock one of the atoms off of the sugar, you will release all the energy that was stored up when it was stacked. When you digest sugar you are basically knocking atoms off of carbon. And knocking hydrogen off of sugar is how you got the initial energy to stack the box in the closet in the misbegotten effort at organization, which subsequently severely damaged your foot. The moral is, never waste energy getting organized.

OK, actually that whole example is a little misleading because electrons, protons, and the like don't seem to be affected by gravity. In the atomic scale of matter the electromagnetic force is more important. But the idea of using gravity to exert force is used in making electricity with hydropower. Water falls due to gravity and turns generators which turns gravitational force into electromagnetic force.

Are you getting the idea that gravity (and the whole idea of forces and energy, are kind of ghostly in properties? For example, gravity appears to be weaker than other forces. Electromagnetic force is stronger than gravity and can lift objects away from gravity. Yet gravity seems to have an affect over great distances whereas magnetism is much more limited in its reach. How can something be weak, yet act over great distance?

Because electromagnetic forces are more powerful than gravity over short distances, gravities influence on the behavior of sub-atomic particles plays no role in determining the internal properties of everyday matter. On the other hand, gravity is the dominant force in shaping trajectory and orbits of astronomical bodies.

Gravity is responsible for causing the Earth and the other planets to orbit the Sun; for causing the Moon to orbit the Earth; for the formation of tides; for natural convection, by which fluid flow occurs under the influence of a heating the interiors of forming stars and planets to very high temperatures; and for solar system, galaxy, stellar formation.

There are a lot of things about gravity that I don't understand. Apparently some people do so if you want you can ask them and maybe they can explain it to you. For example I am told that "the gravitational force is mediated by a massless spin-2 particle called the graviton." What the heck does that mean? I am also told that gravitation is most accurately described by the general theory of relativity proposed by Einstein. Einstein claims that gravity is the result of the curvature of space-time. With all due respect for Einstein, what is he talking about?

ENERGY

If you put a heavy box on the top shelf of the closet, you have created potential energy. That's because gravity is pulling it towards the floor. It's called potential energy because the box isn't really doing anything. But the box has the potential to fall from the shelf severely injuring your foot. There is no need for you to do this experiment. Just take my word for it.

You probably didn't realize that a ripe apple is actually potential energy. You probably thought it was just an apple like Newton did. It wasn't until it fell that he realized that the apple had just realized its potential.

When the box, or apple, falls, it is no longer considered potential anything because it is now actually doing something, namely falling. The movement of the box is called kinetic energy. Kinetic is defined as, "of, or relating to, the movement of physical bodies." I will now use the word in a sentence. My grandchildren are very kinetic. Gravity often causes kinetic energy. Apples usually become kinetic if they are not picked first.

Physicists tell me that the amount of energy released when a box falls is exactly equal to the amount of energy that was required to lift the stupid box onto the shelf in the first place. This is one reason why I don't trust physicists very much. The chaos, damage, and pain created by the falling box obviously exceeded the energy required to lift it. But if you can suspend disbelief for a few minutes, I think I can show you why this idea is extremely important.

Let's suppose that you lift one of your kitchen table chairs onto the table. It now could potentially fall down. Next lift a second chair and place it on top of the first chair. Which chair has more potential to fall? The second one of course, because you had to lift it higher than the first chair, which took more energy. Besides, the higher the stack the more unstable it is. Now lift a third chair on top of the other two. Now what has more energy?

This is a bit of a trick question since you will guess it is the third chair for the reasons just explained. In fact, your wife will probably be the most energized as she realizes the threat to her orderly household! Disregarding your wife, it would be the third chair. Disregarding your wife, however, would not be wise. Perhaps you should do all this when she is not home.

The point is that the more things that are stacked on each other, the more energy the top part of the resultant stack represents. What if we didn't use kitchen chairs but instead stacked shoe boxes? The amount of energy in the stacked shoe boxes would be less, sure. But the third box would still possess the most energy of all the boxes.

What if we stacked nickels? The top nickel would represent the most energy, and the more nickels in the stack the more energy. Even if you stack playing cards the act of stacking would still represent the storing of energy. This may not seem very significant to you because even if a playing card fell on your foot it wouldn't hurt much.

Furthermore, the concept doesn't change when we pick up a tiny particle like a hydrogen atom and stack it onto a carbon atom. It is really no different from stacking a kitchen chair on the table, except that your wife probably won't get all mad about it. You will have still created potential energy. If you stack more than one atom onto the carbon atom, you will have created more energy than if you stack only one. A stack of hydrogen atoms on a stack of carbon atoms is otherwise known as sugar.

Now, if you knock one of the atoms off of the sugar, you will release all the energy that was stored up when it was stacked. When you digest sugar you are basically knocking atoms off of carbon. And knocking hydrogen off of sugar is how you got the initial energy to stack the box in the closet in the misbegotten effort at organization, which subsequently severely damaged your foot. The moral is, never waste energy getting organized.

Anyway, the force with which an apple hits you the head when it falls is a measure, I guess, of the force it took to lift all the water and atoms necessary to make an apple from the ground to the twig on the tree to which it was attached. Of course Newton couldn't have known that in his day because we didn't quite know about atoms and how photosynthesis works yet. He just figured out the easy part.

MASS

Gravity is a force that acts on objects and makes them move towards other objects. If the objects are blocked in some way from moving toward each other it is called potential energy. If I the two objects are allowed to move towards each other the energy is released and it is called kinetic energy.

So it seems obvious that there is some kind of intuitive relationship between gravity, energy, mass, and motion. Almost every person in our modern world has seen the equation produced by Einstein, $E=mc^2$, where E= energy, m= mass, and c^2 represents the speed of light.

It is beyond the scope of this book, and my intellect, to delve deeply into the theory of relativity. However, hopefully you can now begin to see how Galileo's ramp and marbles, Newton's law of universal gravitation, combined with considerations of mass, acceleration, velocity, and energy can make for a stimulating activity and learning experience such as found in part 1 of this book. The body teaches the brain, before the brain can direct the body.

ABOUT THE AUTHORs

Gary McCallister is Emeritus Professor of Biological Science at Colorado Mesa University. He is the author of over sixty scientific papers, mostly in the field of Parasitology, and the award winning author of literally hundreds of popular science articles. He taught numerous subjects in the biological sciences for over forty years at Colorado Mesa University.

Zane McCallister holds a Bachelor's Degree from Colorado Mesa University and is the director of the Grand River Mosquito Control District in Grand Junction, Colorado. He has advanced education in science education and over twenty years in vector control biology. He has held numerous positions of leadership in regional and national organizations in mosquito and vector control associations.

www.ingramcontent.com/pod-product-compliance
Lightning Source LLC
Chambersburg PA
CBHW050815180526
45159CB00004B/1671